There is only one happiness…

YOU

---

this page intentionally left blank

THE 5 SECRET CODES OF HAPPINESS

LIFE IS TEN PERCENT WHAT YOU MAKE IT AND NINETY PERCENT HOW YOU TAKE IT!

WRITTEN BY

ANNE LUSTERIO

The Five Secret Codes of Happiness
Copyright @ 2019 by Anne Lusterio
All rights reserved.

Book design by Anne Lusterio

ISBN: 978-0-578-22181-6

ACKNOWLEDGEMENTS

On May 12, 2018, The Art of Pain, which was the first book I ever wrote, was published. It serves as motivation for me to continue writing and to bring messages to humankind. It is a gateway for my soul, in which I find meaning and purpose on my journey.

I express my special thanks:

- To my one and only Almighty God who is always ready to catch me when I fall. Without you by my side I will be lost in this life. You taught me to appreciate every struggle on my journey.

- To my beloved parents Daniel and Lilia, thank you so much for giving me life. You are my strength.

- To my ever loved brother Daniel Jr., your memories are still very fresh in my mind. I lost you 16 years ago but I remember you always every single day of my life. You inspire me to be a better version of myself.

- To my Godmother Esther, your generosity always reminds me to do things better.

- To Sr. I and Sr. Edward, both of you are blessings to me. I miss our moments every day.

- To my best friend Jeanne Rose, I thank God for having you as my gift. You have shown me the spirit of friendship. You bring so much joy beyond words.

-Anne Lusterio

this page intentionally left blank

"Happiness is a journey...
not a destination."

Ben Sweetland

this page intentionally left blank

TABLE OF CONTENTS

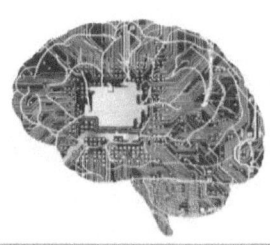

1 **INTRODUCTION**
What You Will Learn from Reading This Book

5 **CODE NO. 1**

37 **CODE NO. 2**

61 **CODE NO. 3**

87 **CODE NO. 4**

115 **CODE NO. 5**

145 **CONCLUSION**
What is the price of happiness?

this page intentionally left blank

THE 5 SECRET CODES OF HAPPINESS

1. Social Connections

2. Kindness and Compassion

3. Reconciliation

4. Mental Habits

5. Gratitude

this page intentionally left blank

Introduction

WHAT YOU WILL LEARN FROM READING THIS BOOK

This book will guide you in how to experience happiness for the rest of your life. All you need to do is follow the codes and you will achieve life's meaning. It is an inner dialogue within your soul. Happiness starts from within your very own self. You are a masterpiece, a complete creation.

What is happiness?

Happiness can be defined in so many ways. One can describe it as focusing on the goodness of the well-being of other people. Aristotle believed that happiness is about living a life of virtue and it can only be judged when looking at your life as a whole.

Introduction

In Buddhism, they teach about calmness and composure during hard times. Some scientists define it as a sense of our life going well, a transitory emotion and a characteristic that we have. They have specified on the first two aspects: life satisfaction and positive effect which combine to form subjective well-being.

Happiness can be observed in behavioral indicators such as facial expressions.

In ancient times, people thought that happiness was the result of **"luck or divine favor"**. That is until the 17th century, when the "happiness revolution" took place. Happiness was declared to be a natural state, a right and one of life's goals to increase pleasure and decrease pain.

It is very normal to feel negative emotions in life. Sometimes our efforts to be happy are frustrated if we encounter such feelings. Actually, it's about finding balance between these perspectives and notions of virtue, to understand what happiness really mean.

Western traditional views on happiness

The happiness of Western people focusses on being more individualistic and high-spirited. They have set internal attributes such as preference and desires for the things that make them happy. They believe that happiness is interpreted as being a **"personal achievement"**.

Eastern traditional views on happiness

While Eastern culture focuses more on being communal and calm, they perceive that each individual is attached by the social relationships they acquire. They don't focus much on personal achievement, but rather on the pursuit of socially desirable obligations. Theirs is more likely a collective self-esteem, which means that individual value relies on the evaluation of others. The culture defines happiness as not being within the individual because for them, personal pleasures can lead to the destruction of social relationships.

Introduction

Is a happy life different from a meaningful one?

Often, we view health, money and comfort as affecting happiness but not meaning. Happiness is often about the present, while meaning encompasses the past, present and future. We look at happiness as receiving and meaning as giving.

**HAPPINESS IS MADE NOT FOUND...
IT ALWAYS BRINGS BEAUTY!**

CODE NO. 1

SOCIAL CONNECTIONS

"Each one of us has an untold story."

this page intentionally left blank

Social connection is very important to happiness. It gives us support during life's challenges, helps us to see our strengths and provides meaning. Research on positive psychology shows that happy people have rich relationships. Talking to friends is one of the most enjoyable activities that provide us with many positive emotions. On the contrary, loneliness is associated with health problems such as hyper inflammation, decreased immune system response, trouble with sleeping and feelings of being excluded by others. These issues create the same effect in our brains as pain.

Pessimism about pursuing happiness

There are three main components that we might not be able to change in our level of happiness, according to Sonja Lyubomirsky:

1.) **We have a genetic set point which accounts for about 50% of our happiness at any given time.**

2.) **Happiness is a personal trait and is mostly non-flexible.** One of the factors is extroversion and neuroticism which are highly linked to happiness and unhappiness.

3.) **Hedonic adaptation eventually adapts to any positive thing that occurs in our life and our happiness will return to its former levels.**
Although hedonic adaptation is proven, we still fail to predict how much and how quickly we will adapt to positive and

negative circumstances. This is called **"impact bias"**. Human beings are very poor judges of what makes us happy or unhappy. We fear break-ups even though people who have experienced them bounce back.

In today's modern society, people are becoming less happy in certain respects, evidenced by our higher divorce rates, lower marital satisfaction, increases in loneliness and having fewer close friends.

According to **John Bowlby**, families become attached to each other based on *three systems: reproductive (sex), caregiving (between parents and babies), and attachment (love and commitment).* These systems create working models in our brains. We think deeply about whether other people are worthy of our trust and how to deal with them.

Code No. 1

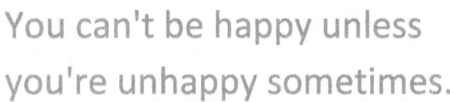

You can't be happy unless you're unhappy sometimes.

-Lauren Oliver, Delirium

Attachment has three styles:

1.) **Secure** - those who are securely attached are loving, warm, and trusting. They have more positive emotions and most of the time they are happy. They are very optimistic and supportive beings.

2.) **Anxious** - people who suffer from anxiety in relationships tend not to feel close enough to others or love others enough. They are often people who have experienced divorce, death or abuse at the hands of their loved ones. They are prone to depression, drug abuse, anxiety and eating disorders.

3.) **Avoidant** - this person avoids closeness, remaining aloof and distant.

Code No. 1

secure attachment

anxious attachment

avoidant attachment

In summary, people with anxious and avoidant attachment styles are considered insecure. However, as they say, there is a cure for everything. We need to think about the positive relationships we have had in the past, or alternatively, we need to cultivate a relationship with someone who has a secure attachment style.

The attachments we experience throughout life are shaped by our early childhood experiences. It has been found that the volume of oxytocin released and used by our brain delivers the quality of life we have. Those with the highest volumes are typically people who are most securely attached. Anxious attached people often experience **"amygdala"** responses to negative feedback. The amygdala is a part of the brain that detects fear and helps us prepare for emergencies. Avoidant attached people on the other hand, have a reduced response to positive feedback. Fearful avoidant people are afraid of being hurt. They would rather be alone than be in pain for the rest of their lives.

One of the ways to combat insecure attachment is to seek a better understanding of our personal character. Communication is crucial when you are in a relationship. Being open to each other can help improve the relationship. As part of human evolution, we are built to connect. As the famous quotes goes **"No man is an island"**. We need someone in our life as much as they need us. It is a mutual relationship of giving and receiving.

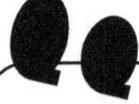

To get the full value of joy, you must have somebody to divide it with.

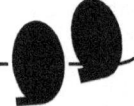

-Mark Twain

What is the Vagus nerve?

The Vagus nerve is a mammalian nerve that starts at the top of our spinal cord and runs downward through the neck muscles, with which we use to nod, make eye contact and speak. It has many key connections to physical functions including oxytocin networks, immune system response and inflammation response. It also correlates the breathing, heart rate and digestive process.

Would you believe that the feeling of being excited and anxious are the same? The only absence for both is in the "breathing". Whenever we feel anxious, we need to take a deep breath. It makes us feel so much better.

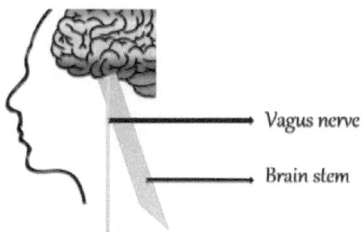

Code No. 1

The Vagus nerve associates the feeling of connection and caring. We tend to respond strongly to empathy and weakly to emotions like pride. It has been found that people who have increased vagal activity, exhibit more positive emotions, stronger relationships, more social support and altruism.

Oxytocin - the love hormone

Oxytocin is a neuropeptide, a series of amino acids that affect the brain and organs. Its levels are increased by touch. In some countries they have a therapy called **"Touch Therapy"**. Human beings have the ability to heal each other by modulating energies, as an alternative to the treatment of specific physical diseases. There is the assumption that illness is an imbalance in a person's energy field and that once the energy levels are balanced, health is restored.

In some cases, touch can be used to communicate emotions, which can be more effective than by voice or by being face to face. Touching someone creates the feeling of reward, safety and of being loved. Being hugged by our romantic partner serves as an antidote in times of stress and worry.

Sadly, in many countries around the world, there are societies that are touch-deprived. Eye contact between men and women are forbidden, especially in public. Culture has a large impact on our well-being. Have you ever noticed that moment when a new born is held by its mother and they gain more energy and instantly heal and bond? It proves that a mother's touch releases oxytocin, especially when breastfeeding, resulting in a quicker recovery.

Since oxytocin is a feel-good hormone, when we experience positive emotions, more oxytocin is released within our bloodstream, which helps reduce stress. This is very helpful in our daily activities. We experience peacefulness and a sense of security. However, sometimes there can be negative effects from hyper oxytocin, which can make us exclude others and focus only on the person closest to us. This could result in us becoming a very committed lover who tends to forget negative social encounters. This could lead to being fooled by someone if we lose our ability to be objective.

Having raised levels of oxytocin helps us cooperate when working in a team. We don't consider our own emotions when making decisions and give greater significance to group agreement. As humans, it is natural to experience feelings of doubts. Whether or not the person we are dealing with is trustworthy, we just go with the flow.

I myself can barely relate to this. Since oxytocin makes me feel good every time I experience encounters with friends, I tend to focus on the bright side and am often blind to the feedback of others. I don't see the flaws of the person involved because I always trust that individual. I often realize too late when I am being taken advantage of. I look at those moments and can say that the same person who informed me about their negative side, was right. I'm the kind of person who, once you became my friend, I trust you already and I can't see nor observe those negative flaws, to the extent that it can cause me harm. Those lessons often cause me to say "no" which is okay. It doesn't mean that they are a bad person. It's just I need to protect myself from the people who are taking advantage of me.

It is important to be aware of who we are and what we want. By drawing a line, people will respect us. Self-awareness is very crucial in life. If our eyes are wide open to the circumstances around us, we are better prepared to react appropriately, which also increases our level of **"Emotional Intelligence"**.

Code No. 1

***THE HORMONE OF CLOSENESS
ROLE OF OXYTOCIN IN RELATIONHIPS***

Our voice is a primal way of connection

We use our voice to communicate. It is a primary tool for conveying messages to others. When we feel sad, worried, angry, excited or happy, we express it loudly. There's a magic of vibration that gives spirit to the words we utter. We can also communicate not only by speaking, but also by singing the rhythm as a display of our emotions. This enables us to convey a more accurate message to the person with whom we are trying to communicate with, as expressed by the axiom: **"The human voice is the organ of the soul"**.

Part of the speaking process is the way we breathe. We need to breathe correctly which means that it should be coming from the diaphragm. Breathing is the foundation of every activity in life. It can help us to relax. If we breathe, it means we are alive.

Code No. 1

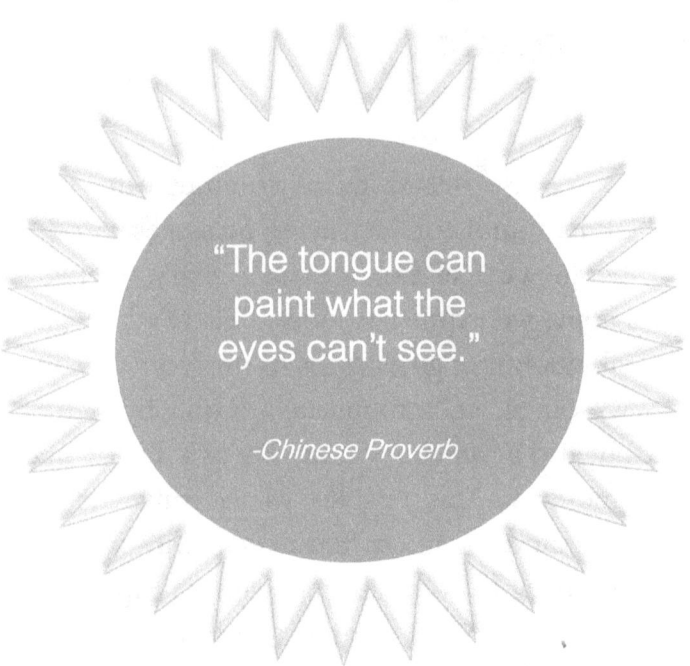

"The tongue can paint what the eyes can't see."

-Chinese Proverb

How to be an active listener

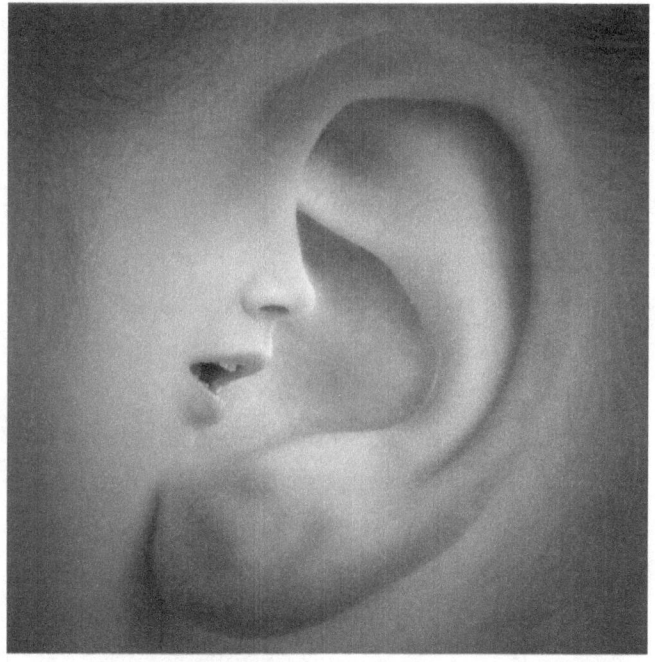

Research has shown that listening to someone increases happiness. It is one of the essential skills a person can have. The quality of our listening power determines our effectiveness in our jobs and in our relationships. Listening serves many purposes. We listen to procure information about something we need. We also listen when we want to understand the message.

someone is trying to convey to us. We want to connect by listening and understanding the subject and to offer support. By showing eagerness about what someone is saying, we express enjoyment at our interaction and we feel good about the conversation. And most of the time we learn from them too. With every bit of discussion, we are learning something. That's the beauty of being an active listener.

When we listen, especially in tough matters, we show empathy which makes us and the other person feel good. Good listeners don't judge and let the other person finish talking without interruption. Despite the fact that even though we focus when listening to the conversation, **we only remember 25% to 50% of what we hear.** Imagine if you are talking to your friends for 30 minutes, you will only remember 15 minutes or less of your discussion. And that is why we need to capture the most important points during the conversation.

Being an active listener requires self-awareness. We need to be consciously aware of what's going on in the conversation. Understanding our own style of communication can be a big factor in being a successful conversationalist. One way to improve our listening skills is to let the other party complete their message without counter arguments while the person is still speaking. We have to stay focused and not get bored. We need to let the other person know that we are listening of what they are saying. We need to be in **"self-awareness"** mode when we are engaged in a conversation. However, there may still be instances when we are talking, that we feel that we are facing a wall brick because the other party shows no reaction to or engagement in the conversation. It is natural to feel bored or frustrated when our message is not being understood by the person we are talking with.

Code No. 1

THE MAGIC OF LISTENING

Romantic relationships

We often misunderstand the context of what really defines romantic relationships. Perhaps we focus more on the physical connection, when actually it's about mutual bonds of love and devotion. Romantic relationships prioritize the well-being and happiness of the partner over their own. It involves compromise and respect and means accepting our partner's flaws.

Code No. 1

There are two types of love:

Passionate love - it is the intense emotions, obsessive thoughts and a desire of union which we have for our romantic partner. Lovesick is another term associated with this type of love.

Compassionate love - it is the emotional attachment and feeling of being in love with our partner. This is the feeling when we feel safe, secure, at peace, comfortable, trusting and close to our partner.

With regard to the above categories, there are two dimensions involved in interpersonal attraction.

Capacity - this evaluates the partner's ability to facilitate our goals and needs. This also correlates with their competencies and the resources they possess.

Willingness - this is the measure of our partner's motivation to facilitate our goals and needs. It entails the sharing of competencies and resources, including their morality and cooperation.

Sometimes we don't need the judgment of others if we want a really successful romantic relationship. It just depends on how we see it. The relationship may be meaningful and enjoyable. But it can also be destructive, unpleasant and hateful and could be the cause of both joy and pain.

Code No. 1

Oftentimes, romantic relationships can be very challenging. People have difficulties in adjusting to sharing their life with someone else. It is human nature to have conflict with one another. Some relationships overcome this conflict and continue to flourish. Others succumb to individual differences. People are wired to change and it may become too tough to maintain the partnership.

The Five Secret Codes of Happiness

Code No. 1

Why do so many relationships fail?

There are nine reasons:

1.) **Unrealistic expectations from yourself** - Look at yourself in the mirror and be honest about it. You can't be a "perfect partner". Putting pressure on yourself is like torture and only destroys you. Let your relationship flow naturally without being forced. Understand your limits and the things "you can" and "you cannot" do.

2.) **Unrealistic expectations from your partner** - In the same way that you perceive yourself, look at how you treat yourself. Most of the time, people get into relationships for the wrong reasons. This is when someone expects that their partner will make them happy. Or otherwise, they look at their partner as if they are their servant. If you have this kind of expectation, be prepared for disappointment unless your partner is so blindly in love with. This type of relationship cannot last.

3.) The need to be right - Conflict and disagreement are always part of this kind of relationship. You know yourself when you are right and wrong. The importance of this is acceptance. Even when you think that you are right and your partner disagrees, let go of it. Surrendering in this instance doesn't mean you are weak. It is more about respecting the view of your partner.

4.) Jealousy and insecurity - Being jealous of your partner means that you don't trust them. On the contrary, they say that "It is an art of loving", but in reality, lack of trust can kill a relationship. A partner who is always jealous tends to make them possessive and creates a very unhealthy relationship. Most of the time, we are afraid of losing our partner which creates a dilemma for us, which can lead to a nightmare of our own creation.

5.) **Self-interest** - Being in a romantic relationship requires sharing. Both parties are willing to sacrifice themselves for the sake of their partner. On the other hand, we will also allow our partner to grow by having their back and allowing them to fight their own battles. Relationships need freedom. You know when to stand and when to back off.

6.) **Being defensive** - This is when you find that whenever your partner has something to say, you always feel the need to defend yourself, even if you are wrong. This is an indication of a toxic relationship. It is important to listen to your partner's wants and needs and respond accordingly without conflict.

7.) Hatred - It is important to forgive your partner and let the past go, if they have transgressed in the past. You don't have to resent them all the time and keep recalling the past. Holding a grudge against your partner is a waste of time and energy. People are imperfect. We make mistakes that we don't mean to. Forgiving your partner can conquer adversity and strengthen your relationship.

8.) You love with condition - This is like "I love you because I need you." When the need is not there you will stop loving the person. Don't love with expectation. Unconditional love is a vital tool for a successful relationship. When you are in love, everything you see is beautiful.

9.) Dishonesty - Communicate with your partner openly and with honesty, especially when it comes to your feelings. Tell them how much they mean to you. Always make them feel your love and presence. Remind them that whatever happens, you will always be there for them.

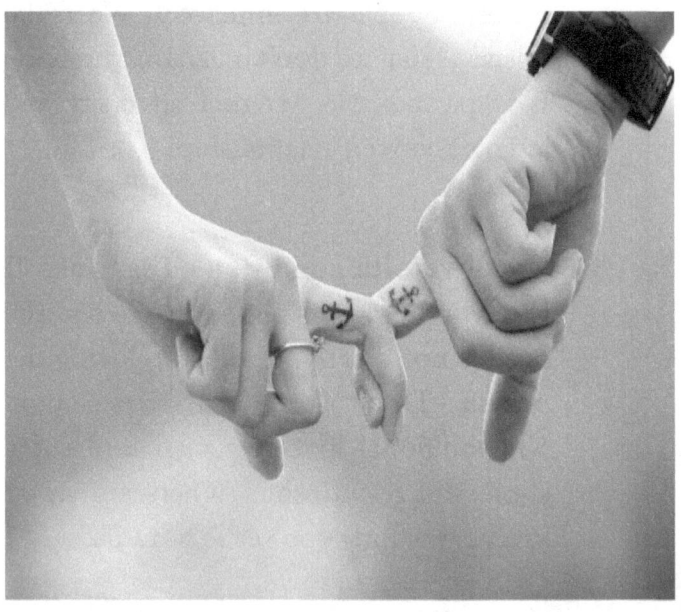

CODE NO. 2

KINDNESS
AND
COMPASSION

"We can heal any kind of wounds if we practice compassion."

this page intentionally left blank

What is compassion?

Compassion is the feeling of witnessing someone else suffering and wanting to help them. Acting on feelings of compassion involves sacrifice. We are to speak for those who cannot speak for themselves. Defend the people who are unable to exercise their rights. It is like living in someone else's body. It is the feeling of joy and peace when we make sure that someone else is at peace and happy as well. We give hope to people every time we show compassion.

Every human being has a **"compassionate instinct"**. Every time we exhibit compassion, some circuits of our brain activate and release oxytocin. It calms our nervous system and slows down our heart rate. We feel so happy every time we help other people. The reward/pleasure centers in our brain are activated.

Code No. 2

What's the difference between compassion and empathy?

Oftentimes we confuse the definition of these two words. **Compassion is the ability to help other people by reducing their suffering.** It is an additional element of empathy and the lifeblood of human morality.

Empathy is to put ourselves in the shoes of others. Having empathic feelings doesn't necessarily mean that we want to help. It dissolves the boundaries between one person and another. It's the antidote to selfishness, although it is not the guiding principle on compassion and kindness.

The Five Secret Codes of Happiness

Empathic feelings whenever we saw this kind of photo.

This is what we called compassion.

There are two types of empathy:

1.) **Emotional empathy** - is the sensation and feeling we get in response to the emotions of others. This entails mirroring their feelings, either when they are stressed or fearful.

2.) **Cognitive empathy** - refers to the understanding of people's emotions which activates areas in the prefrontal cortex that involve language and the processing of semantic content. People who exhibit this type of empathy are highly conscious.

Controlling emotional empathy is a decision-making skill. Holding off on a decision is a sign of strength. We want to make sure that we have enough reason why that particular choice has been made.

Collapse of compassion effect

This is what happens when mass suffering occurs, which has the tendency to make people turn away. In this state, the amount of people in need increases, whereas the value of compassion that people feel towards their fellow human beings, decreases. The reason behind this is that emotions are not triggered by aggregates. Usually this type of individual is skilled in emotional regulation.

Three kinds of compassion according to Buddhism:

1.) **Compassion focused on sentient beings** – These are beings who are born against their freewill. The example of this is poverty and sickness. The person who experiences this type of compassion wants the other person to get them out of their situation.

2.) Compassion focused on phenomena - These are people who suffer a conditioned existence. It is human nature to have the mindset that we are permanent, and as a result we are attached to earthly materials which can lead to our suffering. The individual who has this compassion, understands that all things are impermanent.

3.) Compassion without focus -
Compassion without focus are those people who recognize the emptiness of all things. This type of compassion is very uncommon because it involves both selflessness of phenomena and the selflessness of the individual.

If compassion is the answer to so many of the societal issues that exist today, then how can it be cultivated?

Compassionate people always look out for the welfare of others. And that's the reason why they are able to build connections easily. Being mindful is very important when practicing compassion. It allows us to develop a different relationship with our feelings. We can sort them as if they were clouds floating by. We don't allow our feelings to influence our decisions. People who are compassionate have high levels of emotional intellect. They have a strong sense of morality and tend to be very genuine people.

Code No. 2

Compassion has three dimensions:

1.) **Receiving compassion** - This is the time in our life when we allow other people to touch us. We welcome them with an open heart and we acknowledge their presence and their love for us. Receiving compassion is not selfish behavior. When someone cares for us and is there for us, we also help this person to grow and experience happiness.

2.) **Self-compassion** - When we take care of ourselves, we have the knowledge and experience to give care to other people. Taking care of ourselves includes self-reflection, forgiving ourselves for the mistakes we make and allowing our inner being to grow. We take good care of our diet, we take proper rest and have good sleeping habits.

Through self-care we open the door to discovering our strength and thus enable us to attain proper healing.

3.) Extending compassion - When we extend compassion to other people, we help them alleviate their pain. We help them to be more positive about what life has to offer. We help certain individuals to promote life and make this world a better place to live in.

People affected by war.

Can we train people how to be compassionate?

To practice compassion, we need only to look to our heart and brain which is our temple. Our body is a temple and therefore we need to take care of it, if we are to be compassionate towards others.

Seven compassionate exercises:

1.) **Morning practice** - Wake up with enthusiasm and give thanks that we are able to wake-up and be alive again. We commit ourselves not to waste our time on unnecessary things that will not contribute to our growth. We go out into the world and extend this enlightenment to others in need.

2.) Exercise empathy - Human beings by nature have empathy for others. We only have to develop it and let it out and grow. We try to imagine and put ourselves in the shoes of others when someone is hurt, and think about the effect it will have on them. What can we do to make other people feel relieved? This context makes us become more humane.

3.) Everyone is the same - Now in this scenario, we have to visualize that everyone has feelings. If we're hurt, it means that other beings experience pain too. If we are in need, other people are in need too. We share a common denominator. If we look at it this way, we become more kind and compassionate towards one another.

4.) The feeling of relief from suffering - This is the heart of compassion. When we want to be free from pain, we will feel the essence of our existence. When we begin to let go of the things that we can't control and just focus on the positive side, we will attain the goodness of life. The moment we help others to end their suffering, there's a part of our being that celebrates joy. We feel contentment and allow love to blossom in our hearts.

5.) **Kindness exercise** - Every little thing matters. This exercise can be done in any place and not necessarily on big issues. For example, helping an old person to cross the street. Letting someone go ahead in the grocery cashier line. Baby-sitting for a neighbor for free. Helping walk the dog of an officemate. Simple things that can bring joy to someone else's life. What a great feeling, amazing!

6.) **The people who mistreat us** - It is just "**being human**" when mistreatment by others leads to feelings of revenge, anger and withdrawal. When the feelings of anguish subside, we reflect on the background of the person who mistreated us, wondering what makes a child become that type of person. What circumstances in their past made them act so unjustly? Often, their actions are not about you, but are a result of something they are going through. When we make that realization, we became more kind to that person the next time we encounter them.

7.) Evening habit before going to bed - This is a sort of meditation for self-reflection on what happens in our day. What did we accomplish? Were we able to fulfill what we practiced and committed to during our morning exercise? Will we be able to make a step towards our goals and become more compassionate to the people we have met?

Code No. 2

Prayer time.

Kindness links to happiness

What is kindness?

- By definition it is basically our own ethics, our own beliefs and our own values. **Being kind doesn't mean always saying "yes" to everything, even when it makes us feel uncomfortable.** Other people confuse the notion of being "kind" as being "nice", that when you say no to them, it means that you're not kind.

- If we notice that a friend is struggling with very unhealthy habits and we want to help them by not supporting their bad habits, in the end they may well say something bad about us.

- True kindness is not an easy task. The moment we say no to ourselves, we feel the love of it because we learn to embrace our flaws with an open heart.

Code No. 2

If we experience kindness, it decreases negative emotions and makes us less lonely. It strengthens our immune system. According to studies of people who regularly volunteer, they tend to live longer. Those who give care to other people tend to have longer life expectancy. Additionally, people who spend money on others are happier at the end of the day, than those who spend money only on themselves.

For children, parents are not advised to reward them whenever they show kindness towards other people. In fact, parents should help kids to cultivate kindness, so that it comes from their heart. When they grow up, they will become an inspiration for youth.

Through kindness, we change the way we see ourselves. We became so generous and interconnected to those around us. We start trusting people more, which results in our making more friends. Alternatively, kindness makes us happy which in turn leads us to become kind to others. We find compassion in kindness and we become more understanding of someone else's journey while still respecting our own boundaries and those of others. When we extend our hand to someone, it is a genuine act. We don't do it because we want others to be in our debt or to make ourselves look good in the eyes of others. It is an action without expectation of return.

Do you know that if someone is kind, we find them attractive? It is because our brain makes a connection between kindness and happiness. We tend to want to be with people who make us happy. The reward system in our brain activates when we give something, which makes us feel like we are gaining something in return. This kind of pleasure we encounter motivates us to give in the future.

Code No. 2

Challenges of kindness and compassion

In our daily lives, we encounter different experiences and we tend to be a bit hesitant when offering a helping hand to others. We distance ourselves for our own protection. On the other hand, the world deserves kindness and compassion, so it serves our own interests to help others because it makes the world a better place. Kindness can change the lives of other individuals.

We exist because of love from our creator. We are a product of compassion. We are able to see the light through kindness. This is one of the finest gifts we have as people. A simple act of kindness can touch even the heaviest hearted of individuals. We are giving hope to people in need.

We have to keep in mind that kindness is everything. It is the only thing that responds to every single sorrowful one of us. Kindness associated with compassion is the antidote to the drudgery of our existence as a human being.

One of the tools for fostering happiness is fighting poverty. If we help certain individuals in what they need, it is a great opportunity for us to make a difference in this world. Showing kindness is contagious. When someone sees you helping, the rest of the group will likely do the same. And the state of suffering which the community may be experiencing will turn into happiness and gratitude.

Code No. 2

Who is a hero and what makes them so?

I keep on asking myself these questions. As I reflect, one of the great examples of this is our dear "mother". She is the living hero in our lives. The life of a child is incomplete without the presence of a mother. She carries us for nine months with all the suffering she encounters during the period which she endures without recrimination. The only proof of an unconditional love is seen through her heart. She loves us her children before she sees us without us doing anything for her. Our mother is one of the greatest gifts in life.

A hero risks their personal life. Like us, they are an ordinary person. It can be you or me. They are not special people. They have only the kindliest and bravest of hearts and have the capacity to intervene in any situation. Everyone is called to be a hero. All we have to do is stand and help when emergencies for non-family members occurs. We jump into saving a person without hesitation. We do not do things to promote our ego and social status. We do things because this is how we feel.

We believe that heroism is the right choice and it is for the welfare of someone in need.

> One of the living heroes is our parents, honor them!

Code No. 2

Our superhero mum

Our superhero dad

CODE NO. 3

RECONCILIATION

" The strength of your well-being can be measured by adversity so forgive yourself and forgive others. "

this page intentionally left blank

Conflict is a very common behavior of the human being. Thanks to the gift of reconciliation, we are able to piece the broken pieces together again. Reconciliation is the objective work of humanity. It reigns peace on earth. We will never know how much we can embrace, if we don't experience conflict and following with reconciliation.

Code No. 3

Apology

Reconciliation starts with an apology. Research has shown that when someone apologizes it increases the psychological health and positive emotions for both parties.

Below is a list of how to apologize truthfully:

- When we say sorry, we don't caption it by saying "but", because it is a sign of an excuse. We take full responsibility for our actions. We say "I'm sorry for my behavior earlier and please forgive me."

- A true apology focuses on our own actions and embraces it without even telling the other party that we are also in pain. We don't need to promise anything but instead we do our best not to repeat the same situation.

- We need to listen to the person after we express our apology. We say sorry from the bottom of our heart.

- We have to remember that not all apologies are welcome. The other party may not be ready to hear our apology. Healing needs time.

When an apology is done right, it is easier for the victim to forgive.

Code No. 3

The art of forgiveness

Forgiveness doesn't mean forgetting. It implies the acceptance of negative emotions such as anger, grief, fear and betrayal. It doesn't minimize the offense. We are doing it for ourselves. Forgiveness is a blessing that flows from our heart and brings light within ourselves. It is all about freeing ourselves in order to live a healthier life.

It is also the ability to make peace with the word **"no"**, says Luskin. We feel so upset when circumstances don't meet our expectations. We became unhappy when we continue to keep hold of the past and never let it go. We always have the choice to follow the right path. That way is to embrace the **"Choice to Forgive."** In the beginning it is tough, but if we master it by overcoming its challenge, we can attain peace of mind.

Code No. 3

These are steps to forgiveness:

1.) Make a list of people who have hurt you and who you think are worthy of forgiveness.

2.) Take note of the least painful offence and reflect on how much you suffered and how you feel about the issue.

3.) Once you have decided to forgive, think about the circumstances that led to the offense, including the offender's childhood, past hurts and other things which are putting them under pressure.

4.) Pay attention to whether you feel kinder towards the offender and consider giving them a gift.

5.) At the end of the exercise, you can reframe the experience and find its purpose and deep meaning.

You can repeat this process for the more painful offenses on your list. This practice has proven to decrease anxiety and anger and increase hope, compassion and vitality.

Is forgiveness a sign of being weak?

No. It doesn't require anyone except ourselves. It is something we do for ourselves to lower our psychological distress by getting rid of those negative emotions.

Code No. 3

Why do we need to forgive?

We forgive because we need peace in our lives. It is an act of self-compassion. We have to remember that there are no justified resentments. If we harbor resentment, it gives us an excuse to return to our old selves. Removing blame from your life means never assigning responsibility to anyone you have had problems with in the past. **It is the same as saying "I may not understand why I feel this way, why I am in this state, but I'm willing to say without any guilt that I own it". If we take this as our responsibility, it is one way of saying we will not do it again because we have learned from it.**

No one can define who we are and in parallel we don't have the right to define others. When we stop judging someone and just become an observer, we will know the meaning of inner peace. With that state of being, we will find ourselves free from negative energy and we will be able to live a more fulfilling life. A peaceful person attracts peaceful energy.

I myself am a great believer in God. I have Him in my life and resentment has no place in me.

When we are angry and blame someone for whatever happens in our life, that energy will disempower us. We need to release and let go of that negative energy to attain peace. When we forgive someone, we free ourselves from self-defeating energies.

Code No. 3

Cooperation, an essential tool of a long-lasting relationship

Cooperation has benefits for groups as well as individuals. Everyone should cooperate and achieve the best common welfare for all. It is part of human nature and has a distinctive effect on the brain. It also activates reward processing and pleasure centers. When cooperation breaks down, we feel sad and our amygdala gets activated. Deep within our cerebral cortex is the insula, which activates when we cooperate or compete with each other.

There is a dark side of cooperation for those people who perform "altruistic punishment" against non-cooperators, who experience the same activation in reward-processing areas. In both cases, cooperation or punishment of non-cooperators sustains the social order.

As our Creator gives us different gifts both physical and internal, some qualities and talents are distributed among people with special functions. What is mastered by a certain person is not mastered by someone else. It's like fish which belong in the sea. If we ask them to climb out and stay on land, they will die. Each one of us has a role to play and it is very important to appreciate each individual's existence.

Cooperation between human beings is necessary so that the world will function. We need each other's presence and guidance. There are things that other people may see that we do not. Their contribution is essential for us to weigh our options and make decisions. Each person has their own advantages and disadvantages and affects people at all levels. Cooperation is the basic process of social life. Society cannot exist without cooperation. We will only survive if we act collectively.

It is the act of the wishes of other people. Sometimes human beings find their selfish goals are best served by working together with other people. It involves sacrifice and effort.

There are four elements of cooperation:

- **Goal** - we have a common goal to pursue.

- **Joint activity** - we join forces to achieve a goal.

- **Conscious effort** - we are fully aware of our actions and that it's for the common good of everyone.

- **Restrain over ego-centered drives** - we are not thinking about our own benefit but that of the group as a whole.

Code No. 3

When we cooperate, dopamine levels in our body rise. Dopamine is an organic chemical of the catecholamine family and is also a monoamine neurotransmitter which can make us feel good. Adding people to the equation makes it more fun. There's scientific evidence that if we work as a team and collaborate, it increases our levels of happiness which gives us greater life satisfaction. Sport activities are highly recommended because it can enhance mental wellbeing.

At the same time, we need to ensure that we choose the right peers to associate with. Sometimes our friends cannot be our peers. Friends may subjectively judge us in a biased way but our peers will not. These are the people who share the same goals. Our minds are well connected with our peers. Adding the right people can increase our output in everything we do but adding the wrong people can negatively impact everything we do. We need to surround ourselves with good people and friends, and do things together. This will lead to better results and a more fulfilling life in the long run.

Prosperity in life is expressed by the fulfilment

of diverse human purpose. Social cooperation is an essential tool in achieving prosperity and happiness. Human values are one of the drivers for social cooperation. We work as a team to attain prosperity. With specialization and division of labor, the independent individual becomes a social being. It is only a system based on freedom if every person is equally productive. Free markets permit and enable people to achieve their own goals through mutual cooperation.

Social cooperation is based on free-made contracts. When cooperation is voluntary, productivity is maximized because the individual's free choice is a means to achieving greater prosperity for everyone in society. Any interference with the free market is the interference of the freedom of choice and actions.

Code No. 3

Two types of cooperation:

1.) **Altruism** – is giving without taking. It is more likely about self-sacrifice for the community and humanity as a whole. Nowadays, this kind of quality is hard to find in people. Mostly, everyone has their own motives for helping in expectation of receiving something in return. This self-centered characteristic is very common in human nature. Some individuals donate funds to non-profit organizations for the benefit of those within the community who can't afford a living. Others support a scholarship to sponsor those children whose parents cannot afford to send them to school. In my community, there are doctors who provide free check-ups for those in need. They even give free medicine. But very few are motivated purely by altruistic motives. Most often, people just want to make a living at their own jobs.

2.) **Win-win** – is based on co-working individuals who cooperate on the principles of mutuality and reciprocity. It is the principle, "I will help you and you will help me. Everyone will be happy." They benefit from the collaboration even though there is competition. Win-win ensures that people who like each other cooperate without having to operate against common enemies. This type of relationship is open. No concrete ideas of profits and losses are established in the beginning. It is more likely based on expectations of things we do not already have but which we could possibly gain in the future due to this beneficial exchange. The sociologist commonly describes this as a rational approach that considers people purely as rational beings. It engages the theories of social capital wherein a sustainable exchange occurs when both parties share a mutual confidence. The relationship will be broken only when trust is withdrawn. Therefore, this relationship has an end. It doesn't need to be identical because as individuals, we have different values. in our co-working world today, one could refer to this as an exchange of skills.

The beauty of NOW

This is mindfulness, a non-judgmental awareness of the present reality, our thoughts, our feelings, sensations and environment. We are not thinking about the past or future but we are fully attuned to the **"NOW"**. It is important to stay in the moment. Most of the time, our mind wanders to pleasant, unpleasant and neutral things. About 47% of the time our mind wanders, according to Mr. Matt Killingsworth's research project. It has been shown that at the moment our mind wanders, we become unhappy.

There are a number of activities we can do to help focus a wandering mind. When our mind wanders during meditation, a group of brain areas called "default mode network" activates. Scientists cannot say what it means when our minds start wandering. It seems that our minds are simply carried away as part of our brain's maintenance system when we are not thinking about anything in particular. Our brain regions detecting relevant events start to light up. The moment we focus our attention on our breath, the executive brain network takes over. Over time, when we practice this process, we develop more connection between the self-focused part of the default mode network and brain regions for disengaging attention, which make it easier to shut off that area of the brain when we realize our minds have started wandering.

Being mindful is a means to overcoming this phenomenon. We have the power within us to stop feeling reactive and overwhelmed, if we know how to be mindful. Meditation helps with being more mindful and also doing some sports or by just having "me time", pausing from time to time to reflect. Mindfulness techniques include breathing, sitting and walking meditations. It trains the mind to cultivate a certain state that can make us relax. The famous example of this is called Mindfulness-Based Stress Reduction (MBSR). This particular therapy has been used for people who suffer chronic physical conditions. The main reason to meditate is to become awake and to "**be present**".

Mr. Jon Kabat-Zinn states that we become illusionistic if we say to ourselves often that "we are the stars of our own movie". We just filter everything through the lens of I, ME and MINE. We let our thoughts overpower us rather than experiencing the world through our senses. When we start being aware, we see that it has no limits. We start living our life now rather than being scared of constantly living in anticipation. Mindfulness changes our brain. According to set point theory, our attitude has a bigger effect on our happiness than our external circumstances. Through meditation, our brains are shaped by increasing grey matters in areas related to attention, learning, self-awareness, self-regulation, empathy and compassion. The three brain regions responsible for this are the hippocampus - learning, memory and emotion regulation; the temporoparietal junction and posterior cingulate cortex - empathy; and the cerebellum - emotion regulation.

Code No. 3

There are many benefits to mindfulness. Here are some:

- Mindful partners report greater relationship satisfaction.

- Mindful students participate more.

- It improves our social interactions and makes us feel better about the world and our ability to respond to it.

Code No. 3

> *"MANY PEOPLE ARE ALIVE BUT DON'T TOUCH THE MIRACLE OF BEING ALIVE."*
>
> *-THÍCH NHẤT HẠNH*

CODE NO. 4

MENTAL HABITS

"How you respond in life determines your direction."

this page intentionally left blank

The Five Secret Codes of Happiness

Setting mental habits can promote happiness in life. It leads us in the direction of raising our standards. We become better than we were in our past. We always move forward. We develop the attitude of being resilient. We empower ourselves by conquering our fears and doubts. We focus on quality over quantity. We have the self-respect of every person we meet. We quantify our efforts without judgement. We are not afraid of speaking the truth. We always follow the right path and speak up on behalf of the community.

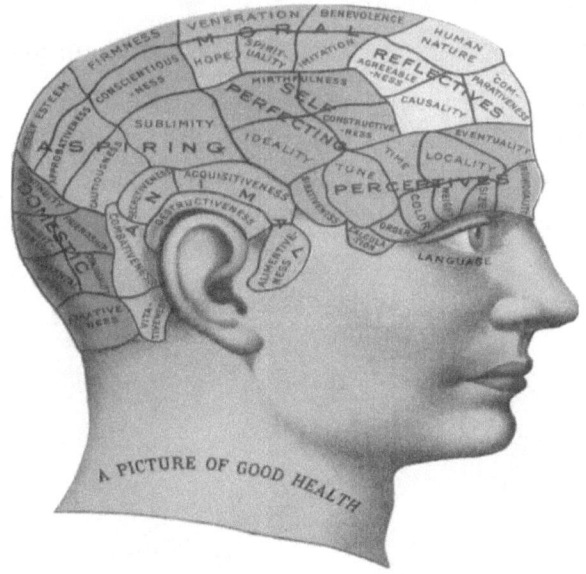

Code No. 4

How to train our brain how to be happy

Meditation shapes the brain. It thickens the insula and prefrontal cortex. People who meditate retain more brain cells while the rest of us who do not practice it, lose 4% of our brain cells as we age. There is scientific evidence that the mind can affect the brain. It strengthens certain areas of the brain and releases different chemicals. Physical brain changes affect our thoughts, emotions and memory.

This two-way pathway makes our minds change the brain and vice versa. One way of changing the brain is to scan the world for positive moments and enjoy them. This path increases our happiness in life.

There are four toxic patterns of thought:

1.) **Perfectionism** - We seek perfection in our life and often feel that we are incompetent. We don't find happiness in every single success we experience. We are not content with our achievements. This attitude we adopt during our childhood when our family affirms us for our intrinsic traits such as intelligence, instead of praising us for our efforts in striving just to get high marks.

2.) **Social Comparison** – We often think of ourselves in competition with someone who we think is superior to us. This attitude can lower our self-confidence and promote insecurity. We don't realize that we are in fact competing against none other than ourselves.

3.) **Materialism** - It is a proven fact that when we book and pay for travel, we create memories. That is the main reason why people become happier when spending on travel compared to spending on material things.

4.) **Maximizing** - This is similar to the definition of perfectionism but differs only on the criteria of usage. Those people who maximize don't content themselves with the good, but rather with only the very best. We need to define the criteria based on our satisfaction. In this world, we are given lots of choices that make us confused about which one to select. We are therefore not content unless we receive the maximum our heart desires. We should take note that once criteria are met, we need to stop looking for more and be happy with what we have.

Being an optimistic person creates a higher subjective sense of well-being and leads to positive emotions which leads to greater future happiness.

After all, life is not perfect.

Comparison through social media.

Code No. 4

The Role of Flow in setting goals

We first need to define the meaning of flow. It is a rewarding state of mind that comes when we are intensely engaged in an activity. Because we are so focused, we lose track of time and therefore forget completely about ourselves and the environment around us. In this state, we tend to be more creative and productive which leads to greater satisfaction.

In order for us to experience flow, we need to set clear goals and ensure that our skills correlate with the challenge we are about to face. In addition to that, we need a place where we can concentrate. We need to check every now and then whether we are in line with our goals.

Aside from experiencing flow, we may encounter a state of "boredom" which we identify as "low stress" where we are trying to focus but we cannot. Another state is **"frazzle"** where we experience being stressed, resulting in poor performance because we are distracted by negative emotions.

We need to set an example to help students experience **"flow"**. In a recent study, most students were found to be obsessed with grades rather than with learning. **Almost 50% of students are bored everyday according to research conducted in 2009.** Because students are aiming for high grades rather than focusing on learning, cheating, depression and drug abuse are on the rise. One method to encourage flow among students is to ask them to pick their own tasks and let them handle them at their own pace. This means that they have less distractions, have more time to focus and therefore are able to experience more positive emotions.

In most cases, flow happens when we are alone. Sometimes it occurs in groups too if there is balance and freedom to allow the group to perform at its best level. When each member of the group is familiar with each other, they perform better, enabling them to achieve flow. This encourages cooperation, where every member of the group will listen and not contradict each other due to ego or hidden agendas.

Code No. 4

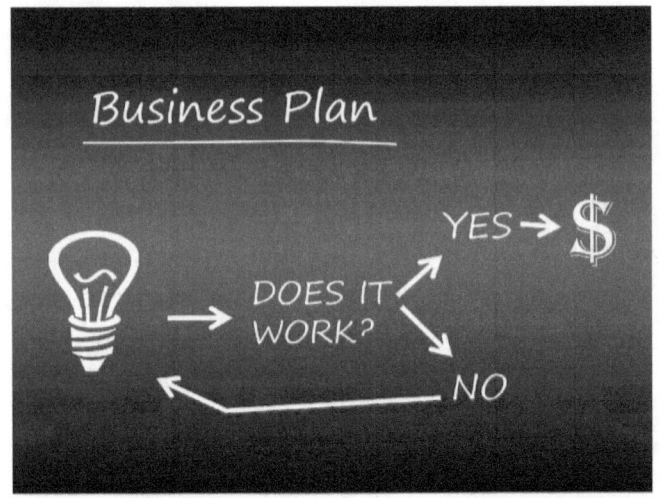

The magic of flow.

Fostering happiness through our goals

When we have a goal, we have a purpose for living. We have the meaning for existence. We don't live merely to breathe. We live because we have a sense of commitment to ourselves.

Goals have two characteristics:

1.) **Intrinsic goal** - this involves psychological needs around autonomy, competence and connection to others. It creates a sense of belonging, especially from the experience of love from the relationships we have. This is the deeper goal that merely focuses on our feelings and how content we feel with our decisions. This goal is a long-term commitment.

Code No. 4

2.) **Extrinsic goal** - this comprises elements such as fame, power, money and all other material possessions. This type of goal is very temporal and can only satisfy our physical needs. This goal focuses on the acceptance of other people. Their approval gives meaning to our life. We place importance on the judgement of other people. We are not true to ourselves because we are always worried that we will not meet their expectations. This goal is very temporary and short-term. This goal provides very little satisfaction.

Intrinsic goal

Extrinsic goal

Code No. 4

Finding our fit

We seek activities that boost our level of happiness. We do a lot of experimenting in order to find out what is really suitable for us. We ask our friends for suggestions and we weigh it by going with them just to measure if it is really for us. We know immediately if a certain program is for us.

There are five drivers of finding our fit:

1.) **Motivation and effort** - This is the measurement of how greatly we are motivated and how much effort we exert just to find out our "why's". It is indeed an amazing feeling when we are able to boost our levels of satisfaction.

2.) **Efficacy beliefs** - This is the power of mind over matter. We need to tell ourselves that we really can do it. We become unstoppable in the pursuant of our goals. The struggles that we face are nothing compared to the reward that we receive.

3.) **Baseline affective state** - In this area, we ask ourselves **"how happy I am?"** Everything starts from ourselves. If we begin things with happiness, our chances of success are greatly increased.

4.) **Social support** - Friends, family and colleagues are the main contributor of happiness in our existence. They are a part of our day-to-day experiences. These are the people who encourage us. They are the ones who believe in us. They catch us when we fall and help us to be resilient. They stand with us without judgment.

5.) **Demographics** - These factors comprise our age, sex, religion and culture. Demographics play a large part in finding our fit, even though in this era, the stress is on pursuing "equality" for everyone. No gender, status, color nor religion can divide us. However, despite the goal of the community, we still cannot convince certain individuals to overcome their perceptions about people outside of their own communities. They still have this feeling of ambiguity that leads them to be racist. They may experience insecurity and thus withdraw from that particular group.

Based on the study of the above list, evidence suggests that when we counter these factors, it will make a big difference in finding our fit. It fosters our happiness at the highest level. By motivating people, it can increase their levels of oxytocin because they expect benefits. Same also when people exert efforts to get better results. In a baseline effective state, happiness activity works for people who are mildly depressed, not people who are happy or severely depressed. Social support works better after reading testimonials of people who have endured traumatic experiences and finding light along their journey. **When we talk about demographics, "*Westerners*" tend to get more benefit out of happiness practices than other cultures.**

Code No. 4

Finding our fit.

"In order to find the right person, we must become the right person."

Five ways to diagnose our personal activity fit:

1.) **It should be natural** - There is no other credit to our actions than doing it with open hearts. It is not beneficial to engage in activities against our will. It is necessary that we feel at ease with the activities we are engaged in.

2.) **It should be enjoyable** - When we participate in events, we must ensure that we will enjoy it. It makes no sense to engage in activities that we are not in the mood for. Events are supposed to be fun in nature. We should be enthusiastic about activities that we are engaged in because we find them interesting and relevant.

3.) **It should be valuable** – The activities we engage in need to be measured by the value they add to our journey. We must ask ourselves "Is this worth doing?". We have to believe it that it brings meaning to our path.

4.) **It shouldn't be done out of guilt** - Whenever it's necessary we must say NO. We will do things because we want to, not out of guilt or shame. Remember, we are talking about none other than "ourselves". Therefore, we must consider our feelings and how it can contribute to our joy.

5.) **It should be based on our choice** - We are liable for every decision we make. We must not feel that we are being forced by our circumstances or social pressure. We do things that make us happy, not miserable. We participate because we feel comfortable not because we tell ourselves that "I don't have an option but to do this". That is really unacceptable and you owe that to yourself.

The narrative magic

What is a narrative? It is a structure we use to make sense of our lives. It is a connected event, a story for every individual.

There are two main types:

1.) **Micro-narrative** - This is the structure of the daily struggles and challenges of every human being. Through this component, we can be led to easily make this world a better place in which to live. We require this so that we will understand our fellow human beings - their culture and their way of living. In this generation, fully understanding the meaning of the micro-narrative can have a positive effect on the world. We support this goal which can rise above being heard in the overcrowded state.

2.) **Meta-narrative** - This is a broader story of oneself and our journey in life. This will include conflicts and turning points in life. The main structure of this is suffering, compassion and forgiveness. Oftentimes we watch videos go viral in social media. Like myself, I'm an avid fan of TedEx. I usually watch videos on YouTube. I love hearing about success stories.

As research shows, people who are involved in narratives experience higher levels of well-being over time. This correlates with the idea of having "more possible selves", as well as different stories and identities.

Sharing stories and identities.

Code No. 4

Key factors of creating effective mental habits

- **In every action, there is a corresponding reaction** - It is very important for us to focus on our goals. If we want something new in our life, we need to start creating a new version of our old self. Stop procrastinating and continue to move forward.

- **The power of choice** - In every decision of in our life, we always have the ability to choose. As a human being, we are responsible for our journey. We cannot blame someone else for our shortcomings. It is us who decides our future. When we choose something, we need to own it. The moment we separate from our parents' home; we are in-charge of our path. We should be parenting ourselves. We should not negotiate our decisions like we used to when we were living with our parents.

- **Always look at the brighter side of the equation** - Negative emotions are always a part of us but we don't allow it to overpower us. We need to combat it always by thinking about the impact they have on our lives. Negative emotions can lead to fatal illness. It releases large amounts of a toxic chemical called cortisol.

- **The happy person in nature** - As I mentioned earlier, social connection contributes largely to fostering high levels of happiness. We need to be connected to our fellow human beings. We need to show eagerness to listen to other people's stories. With that scenario, they will feel safe and will therefore open their hearts to trust us. They will create intimacy with us.

- **Eat the right food** - Our body is the result of the food we eat. We need to give importance to our health. Remember,

Code No. 4

health is wealth. No matter how rich you are, if you are sick, your money will not replace your health. It is our obligation to follow a balanced diet. We need to think about the plants and animals that gave their life so we can live for tomorrow. We should not waste food too because so many people are starving to death and we are so lucky to have food to eat.

- **Have enough sleep** - We are all aware that because of the presence of social media, we are depriving ourselves of valuable sleeping time. Our brain is subconsciously working and that makes it harder for us to fall sleep. We are getting less hours of sleep. We prioritize nonsense things. We trade our sleeping time for unnecessary and trivial things.

- **Do Exercise** - Your body is like a car. You need to start the engine and drive. Human beings are wired for movement. As we age, our body deteriorates. It loses power and to sustain that, we need to do exercise on a daily basis.

- **Personal hygiene** - We need to take care of our body by keeping it clean and tidy. Ensure that we are not emitting any undesirable smells because it also distracts other people. Personal hygiene attracts people and it affects the first impression we make. They will treat you better if you are physically neat and presentable.

- **Writing notes** - We don't practice writing notes these days because we are busy chatting to our friends on social media. We tend to forget our "me" time. When we write, our subconscious mind activates. It gives us more clarity on our goals when we put it into words.

- **Continuous education** - There is no master in this world who doesn't tell themselves every single day that there are still so many things they need to learn. Reading books, attending seminars and undergoing training is the gateway to knowledge. We will become the master of skills if we keep sharpening the saw. It will change our life for the better.

Mental habits.

CODE NO. 5

GRATITUDE

"A grateful heart creates an infinite door to happiness."

this page intentionally left blank

Gratitude is a feeling of thankfulness from something we have received. It happens when other people give us thanks for something good we have done for them. Gratitude arises from cooperation. Reciprocal altruism is driven by gratitude. It helps people connect to something bigger than them as individuals. It is an emotion we feel in response to a gift. It leads us to be more generous and self-giving.

There are two components of gratitude according to Dr. Robert Emmons:

1) **First is the affirmation of goodness about life** - There are gifts we receive in this world from which we have benefited. They provide us with reasons to live.

2) **Second is the recognition that the sources of this goodness are outside of ourselves** - We acknowledge the presence of others. We become dependent on one another. People give life so that we can have life. In this stage, we recognize people who matter the most to us.

Why is gratitude so powerful?

A simple act that is appreciated by someone who simply tells us **"thank you"** has been scientifically proven to deepen our connection to others. Grateful people experience more happiness and are pro-socially attached. It not only helps us to see more of the good in life but also increases the benefits we get out of the experience. It helps us to remember and reminisce about positive experiences and results in our giving these experiences more focus and importance.

Gratitude is a part of our daily life. It is a character strength. It is also a part of our religion. Expressing gratitude towards greater well-being increases our mental and physical health. Gratitude is a very powerful tool for strengthening interpersonal relationships. People who express gratitude towards their partner are more willing to forgive and be less narcissistic. Giving thanks to those who have helped us, strengthens our relationships and promotes the formation and maintenance of life satisfaction.

Code No. 5

Schools have practices which include activities for students such as delivering letters to someone they want to thank, like parents, sisters, brothers, friends, or spouses etc. It has been proven that right after such tasks are performed, their happiness levels increase. It has direct long-lasting effects. Thus, the more grateful we are, the happier our lives will be.

The Five Secret Codes of Happiness

Code No. 5

The three benefits of gratitude by Dr. Robert Emmons:

1.) **Psychological** - People who are always grateful experience more positive emotions and greater pleasure which makes them more optimistic, energetic and happy. Gratitude reduces the frequency and duration of depression.

2.) **Physical** - Grateful people have strong immune systems. They don't entertain dramas, pain and hurts. Instead they take care of their health like exercising regularly, getting enough sleep and eating good food.

3.) **Social** - Grateful people are more helpful, generous, compassionate and forgiving.

In all cases, gratitude discourages us from taking things for granted. It boosts our self-worth as a person. The positive effect of gratitude helps us to move on from past stresses and prevents us from experiencing negative emotions like envy, resentment and regret.

Gratitude shows us that we are not always in full control or self-sufficient. We don't always get what we deserve in life, instead we get more. Sometimes it challenges our self-serving bias in the way we take credit for the good things that happen to us and blame outside forces for the bad things happen. Gratitude can be cultivated by counting each and every single blessing we have in our lives.

In contrast, we cannot be happy all of the time and gratitude is a very powerful tool when life is tough. Our negative perspective will turn into a positive one. Rather than complaining every time something happens to us that we don't think we deserve, just look on the bright side. People who have illnesses value life. Every morning that they wake up gives them meaning and they are so grateful for another day to live. Gratitude indeed is a choice.

Code No. 5

According to Emiliana Simon-Thomas, gratitude is the **FIND, REMIND** and **BIND EMOTION**. It helps us find people to form a relationship with, reminds us about their good qualities and builds a closer bond with them. Our minds notice good things and interprets situations positively. When we feel grateful, it's an indication that other people are acting properly towards us. Thus, it encourages us to return the kindness and express gratitude in return ourselves, by reinforcing our moral behavior. We love grateful people because they bring so much positive energy and are likely to help us. They like to work with others even if sometimes it costs themselves.

The division of labor in a relationship

Most partners, whether in marriage or in friendship, struggle over the division of chores. If one party feels it is unfair, they are discontented and thus consider separating from their partner. Aside from this issue, it is the **"lack of gratitude"** that we oftentimes feel. Not all partners are the same. Some are just leaving the responsibilities of home life like cleaning, cooking and washing to the other partner. Every time, this partner expects the other to do it and therefore they do not consider it as a **"gift"** that requires gratitude.

We must be aware of it. When we deal with our partners or roommates, we must be sensitive to these issues. They are not our slaves - we need to help each other. And in some cases, if they do us a favor, always remember to give them thanks. In this way, they are fully aware that we value their efforts and the same dynamic will apply to the things we do for them.

Code No. 5

Challenges of gratitude

Practicing gratitude results in tension when dealing with our daily habits and personality. It goes against individualism and our sense of entitlement. On the other hand, we reserve our gratitude because we worry it will make us complacent or over accommodating. In my culture, we experience a sense of indebtedness to acts that never end, no matter how many generations have passed. Sometimes the people we are indebted to feel entitled to abuse us because of the thought that we owe them so much. It creates a sense that even the cost of our life is not enough to repay the debt. This is a very sad culture but is the reality of daily life. It has been handed down from generation to generation. It becomes an obligation to the extent that when we cannot satisfy their wants and needs, they tell us "how shameful" we are, that we have "no integrity at all". These words hurt us to the bone but we swallow it because we don't have the capacity to return the sentiment. We just cry out loud and pour our tears into our hearts.

Code No. 5

Myths about gratitude

According to Robert Emmons and Brene Brown, there are four myths of gratitude:

1.) **It is another form of positive thinking** - It is a feeling of being in a commitment to be grateful for all the things that happen in our life. It is a state of being that transcends circumstance either good or bad.

2.) **It strips people of initiative and leads to complacency** - Since gratitude deserves a reward, people expect something for everything they do for you. Just remember that genuine gratitude is selfless and doesn't seek to be reciprocated. Gratitude on its own is a reward. It benefits us mentally and spiritually.

3.) **It is impossible to be grateful in the midst of suffering** - Whenever I experience hardship in my life, I look around and notice people who are suffering more than I do. From that moment, I start giving importance to the life I have and say to myself that I'm still very lucky and blessed. Suffering draws me closer to my Creator. It deepens my personal relationship with him.

4.) **Too much gratitude shows** weakness - As human beings we are often vulnerable. Oftentimes, we are scared to give because of the fear that others may take advantage of us. It takes enough courage to perform such heroic acts. Vulnerability is a part of happiness. We should focus on gratitude and not fear.

Code No. 5

In the midst of pain lies a grateful heart.

The barriers to gratitude:

1.) **Adaptation** - According to Tom Gilovich, people who prioritize events such as purchasing live vacations over material things have higher rates of happiness over time. People who strive more create deeper social connections and thus feel greater satisfaction in life. It also empowers ourselves to be more generous. It doesn't matter if our experiences have some negative impact, it still it brings a source of gratification which is a pathway to self-growth.

2.) **Dwelling on negatives** - Because human beings are so wary and scared of failure, we ignore the benefits of obstacles. People always complains about everything. It prevents us from appreciating the factors that contribute to success.

Code No. 5

3.) **We perceive shared burdens as disproportionally affecting ourselves** - This type of character leads us to resentment and makes us less productive.

Vacation vs. Shopping
Your choice as individual promotes happiness.

Code No. 5

There are six habits of grateful people according to Jeremy Adam Smith:

1.) **Thinking about death and loss** - When we lose someone we love, we think that it is the end of the world. However, grateful people focus on the positive side. It makes us appreciate more and value the presence of those people we love every single day.

2.) **Stop and smell the roses and take delight in rituals** - When we pause for a while, that is the time when we start realizing our own self-worth. It is like recharging ourselves from the exhaustion caused by a complex world.

3.) **See life as a gift** - When we have this perspective, we tend to be more generous. We focus on the welfare of others. We give thanks always every morning when we woke up and make our day as productive as if it is being our last day of living.

4.) **Grateful people activate their biological system of trust, affection, pleasure and reward resulting in greater happiness** - This positivity will reflect in their aura and can be a magnet to those around them. This behavior becomes so addictive because it releases oxytocin in our brain.

5.) **We are specific about the things we are grateful for** - When we say things in particular, we become more authentic. They understand that the givers intention has costs and they therefore value them the most.

6.) **Even if we experience adversity, we are still thankful** - Highly grateful human beings see their trials as a positive thing because without it, their life can be meaningless. They will not strive for their own betterment. There is no challenge and excitement. Deep down in their heart, they believe that everything happens for a reason. Whatever reason that is, they only put their trust in God.

The power of gratitude in a romantic relationship

We value each other's presence when we are thankful. It inspires us to be a better person every day. It encourages us to do something great for our partner that results in a healthier relationship. We don't take for granted our partners' important life events, such as their birthdays or anniversaries and oftentimes we give them surprises. It lifts their hearts and shows how genuine your feelings are towards them.

When we are grateful, we create a cycle of generosity. We listen to the needs of our partners and are sensitive enough that we can almost read their minds. The bond of connection is much deeper.

Code No. 5

Five ways to cultivate gratitude at work according to Jeremy Smith:

We all know that work is a paradox. We want to be the center of attention with regards to our boss and to our colleagues. Gratitude is 60% missing from the working environment because of misconceptions that everyone gets paid to do a job. We observe gratitude as a sign of weakness. We don't acknowledge the fact that we need the help of others. If we could only realize that gratitude can actually make us feel respected and enhance our sense of accomplishment and purpose in life. When we help our officemates, we build trust and make ourselves worthy.

- **It must start at the top. Employees need to hear the words "thank you"** - It serves as a fuel engine and helps them to perform better.

When the employer gives importance to the employee, they will feel safe and secure resulting in their devoting all their effort and commitment to them.

- **Thank people who never get thanked** - Bosses, employers and those in high positions don't realize the importance of this. It is very crucial to express gratitude to those employees who never receive appreciation. The office boy who prepares our daily coffee and brings our newspaper or performs other errands we require. The cleaners who mop the floors, who clean tables, tidy up messes and even clean the washrooms. The operators and laborers who make the products to we sell. Without those people, we will be nothing. Companies will not operate and be successful without their presence.

- **Focus on the quality of gratitude** - Don't give thanks in an exaggerated manner. It will detract from the authenticity of our gratitude; it will become gratitude fatigue. Be specific about why you are grateful. It means that we give genuine appreciation for their actions.

- **Let people express the gratitude in their own way** - Every individual has their own style of expressing how grateful they are. When a boss increases his subordinate's salary, as a sign of being grateful, he will in turn exert more effort in his job and in coming in on time, if not earlier or occasionally leaving late when needed. Other ways to show gratitude is to throw them a dinner. There are various kinds of opportunities for gratitude. Not everyone likes to be thanked or likes to say thank you, especially in public. In this case, a personal thank you note or card would suffice.

- **Use gratitude to help the team get through a crisis and see the positive** - In this practice, the team will build a psychological immune system that can help us when we fall. It is scientifically proven that companies who practice gratitude are more resilient to stress. Employees see the moments of adversity as a gain. It gives them the tools they need to face deal with similar problems in the future.

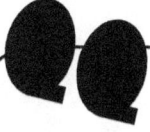

> You cannot do a kindness too soon because you never know how soon it will be too late.

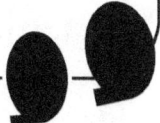

- Ralph Waldo Emerson

Code No. 5

The wonder of awe

Awe is something we feel when we encounter a super power above us. We often call it the Creator, our God. When we start questioning how human beings originated apart from the views of science view or some other miraculous things that cannot be answered by any intelligent person. We experience so many incidents that we habitually call "a **blessing in disguise**". Those instances when we believe that our guardian angel has saved us from danger, cultivates our spiritual belief and deepens our relationship with our Creator.

According to studies, it has been proven that spiritual people are happier and less depressed because they are part of strong communities and they experience more awe due to their belief. People who have strong faith spend more time volunteering and extend help to their fellow beings. They choose to buy experiences over material things. They value memories rather than the physical possessions they have. In addition to the research conducted by Stanford University, experiencing more positive emotions on a daily basis is linked to lower levels of Cytokine Interleukin, a marker for inflammation that leads to dreadful diseases.

Awe can make people generous. We tend to give because we appreciate what we have as blessings given to us by the Creator, our God. We care and therefore share the things what we have. We become more sensitive to the needs of other people. We give because that is what our hearts desire.

Code No. 5

The beauty of Awe.

Conclusion

WHAT IS THE PRICE OF HAPPINESS?

To end this statement, we need to crack the code of happiness and apply it to ourselves first. We are the creator of our happiness. We view the world through our own receptors. If we want to be happy, we need to start changing NOW. Our time is very limited. Tomorrow will never be a promise to each one of us. Change is hard but it's possible. We need to be better every day. Suffering is always a part of our journey. We should look on it as a blessing and start appreciating it.

Being happy means to have direction in life that is socially connected with the people we love. Happiness means showing kindness, compassion and reconciliation with the people

Conclusion

who mean the most to you, by telling them how happy and blessed you are in being part of their journey.

The reward of happiness is experiencing everlasting peace and contentment in life. How do we attain that? By simply following the **5 SECRET CODES OF HAPPINESS.**

"Every little thing that matters to us contains happiness."

"Happiness is a choice you design on your journey, make it a habit."

BIBLIOGRAPHY

- **Bowlby, John.** (1969). *The attachment theory.*
- **Catalino, Lahnna.** (2014). *A better way to pursue happiness.*
- **Emmons, Robert.** (2009). *Gratitude.*
- **Gilovich, Tom.** (2017). *How to Overcome the Biggest Obstacle to Gratitude.*
- **Goleman, Daniel.** (2013). *Focus: The hidden driver of excellence.*
- **Hanson, Rick.** (2013). *Hardwiring Happiness.*
- **Kabat-Zinn, Jon.** (2013). *Using the Wisdom of Your Body and Mind to Face Stress, Pain, and Illness.*
- **Keltner, Dacher.** (1999). *Social functions of emotions at multiple levels of analysis.*
- **Killingsworth, Matthew.** (2010). *A wandering mind is an unhappy mind.*

Bibliography

- **Luskin, Frederic.** (2001). *Forgive for Good.*
- **Lyubomirsky, Sonja.** (2008). *Happiness set point.*
- **McMahon, Darrin.** (2006). *Happiness the hard way.*
- **Simon-Thomas, Emiliana.** (2012). *Social Cognitive and Affective Neuroscience.*
- **Smith, Jeremy Adam.** (2013). *Six Habits of Highly Grateful People.*
- **Zak, Paul.** (2013). *How Stories Change the Brain.*
- **Zimbardo, Paul.** (2015). *Culture and Psychology.*

Other book of *Anne Lusterio*

THE ART OF PAIN

The book is about…
- the beauty of pain
- the power of prayers
- self-image
- your own standard
- life's approach
- self-confidence
- miracle
- comfort zone

ISBN   13: 978-154476531

this page intentionally left blank

www.ingramcontent.com/pod-product-compliance
Lightning Source LLC
Chambersburg PA
CBHW021952290426
44108CB00012B/1043